COMMON PLANTS OF THE EAST BAY

COMMON PLANTS OF THE EAST BAY:

a guide to common non-woody plants found in East Bay communities

Text and illustrations by

Lisa K. Wagner

Appletree Press
Berkeley, California

for Herbert G. Baker
who inspired my interest in weeds

Library of Congress Catalog Card Number: 83-71995

International Standard Book Number: 0-9611956-0-6

Published by

Appletree Press
2644 Dwight Way #6
Berkeley, California 94704

Printed in the United States of America

Table of Contents

Introduction

Interesting plants are not restricted to undisturbed areas and nature reserves; they appear along sidewalk edges, on roadsides, in vacant lots, and in garden beds. Many of the species you will find in these places have world-wide distributions and are known as cosmopolitan weeds. Some have traveled from areas of similar climate, such as the Mediterranean regions of Europe, South Africa, and Chile, and do well in California. Others are garden escapes which have become established in localized areas under favorable conditions. Still others are native species that are able to grow in frequently disturbed habitats and persist in urban habitats as well as in their natural habitats. This guide covers many of the common herbaceous (non-woody) plants you will find in the East Bay.

Each species is illustrated and briefly described. The distinguishing features of each species (habit, flowers, fruits, leaves, habitat) will help confirm your identifications. Species are grouped by flower color, beginning with white, progressing through yellow, orange, pink to red, blue to violet, and ending with brown to green. The glossary on pages 57 and 58 includes any technical terms used and pages *ix* and *x* illustrate flower and plant parts. If you would like to learn more about local plants after using this book, additional references are listed on page 61.

Parts of a Plant

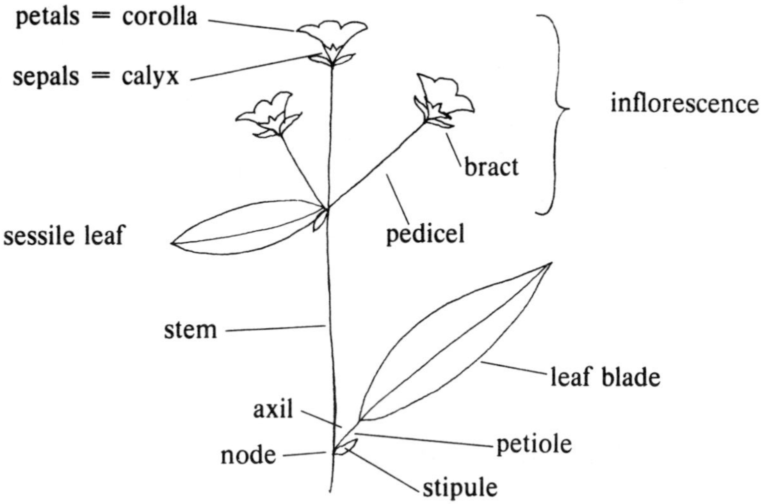

Parts of a Flower

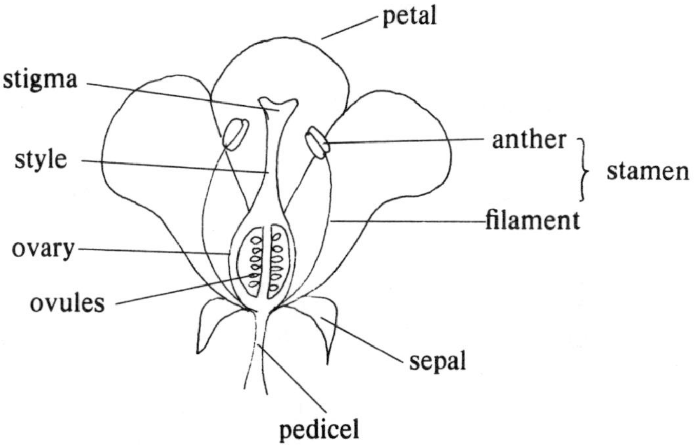

Sunflower family (*Compositae*) Inflorescence

Sunflower head:
top view

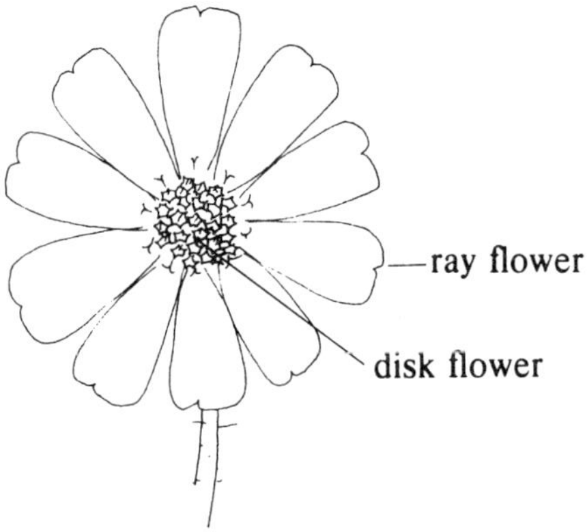

Cross-section of head

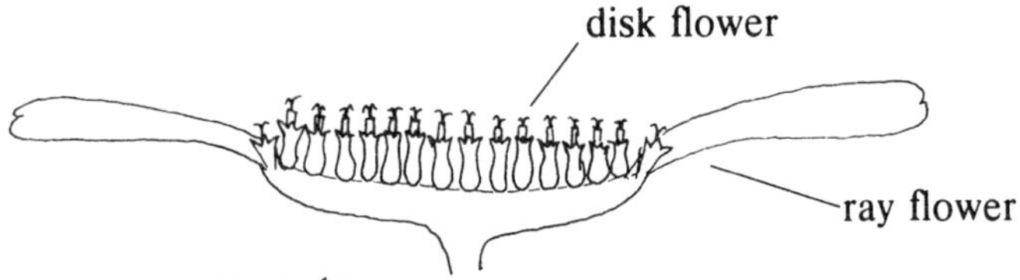

Disk flower

tubular corolla

Ray flower

strap-shaped corolla

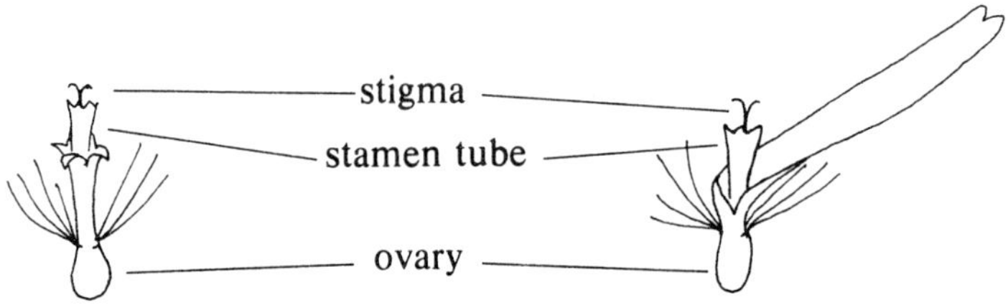

Bellis perennis

English Daisy

Perennial, introduced from Europe

Habit: scapes (flowering stems) 5-20 cm. high, leaves in basal rosettes

Flowers: white to pink ray flowers, yellow disk flowers; head borne singly on leafless stems (all year)

Leaves: broadly oblanceolate to obovate in flat basal rosettes

Fruit: achene

Habitat: lawns and other grassy places

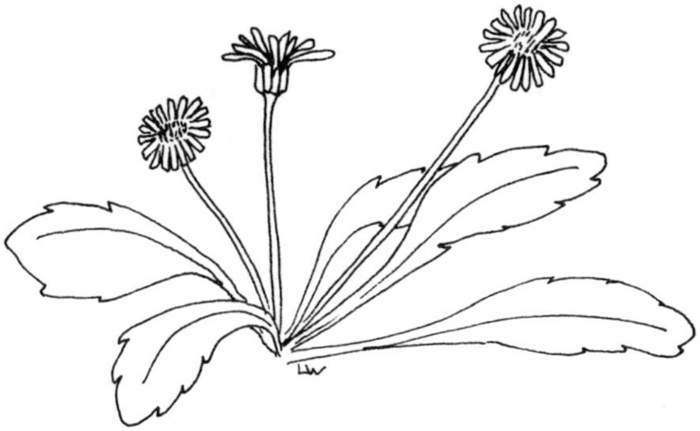

The genus name is from the Latin *bellus* for *pretty.* The flat basal rosette of leaves and the vigorous vegetative spread from leaf axils make this species a successful lawn weed. In full sun, the leaves grow prostrate, pressing down on the surrounding soil and vegetation; in shadier areas, the leaves tend to bend upwards. The flower heads close at night and open in the morning after an hour or so of sun.

Compositae (Daisy or Sunflower family)

Lobularia maritima

Sweet Alyssum

Perennial, introduced from the Mediterranean region

Habit: much branched, .5-2.5 dm. high

Flowers: white or sometimes purplish, 4-petaled, sweet- smelling, in dense heads (all year)

Leaves: narrow, entire

Fruit: oval silicles approximately 2 mm long

Habitat: common garden escape in waste places, along sidewalks

This species is abundantly naturalized in California in coastal areas. Its affinity for a maritime climate may be the source of the specific epithet, *maritima.* The development of a fruit from every flower indicate that this species is primarily self-pollinated.

Cruciferae (Mustard family)

Trifolium repens
White Clover

Perennial, introduced from Europe

Habit: creeping, rooting at nodes, stems 1-3 dm. long

Flowers: white to rose, in heads (all year)

Leaves: 3 leaflets, each obovate; some plants with light- colored leaf markings

Fruits: pods included within a persistent calyx

Habitat: lawns, fields

This species spreads vigorously by means of its creeping stems. The roots, like those of most legumes, harbor nitrogen-fixing bacteria; this increases the amount of nitrogen available to the plant. Dried flower heads, when steeped in boiling water, make a pleasant tea. The flowers are also attractive to bees and make good honey.

Leguminosae (Pea family)

Heracleum lanatum

Cow-parsnip

Biennial to perennial, native

Habit: erect, stout, hairy stems, 1-3 dm. high

Flowers: small, white, forming a large compound umbel that is flat on top (Mar.-Sept.)

Leaves: large, divided into 3 maple-leaf shaped leaflets, with sheathing petioles

Fruits: 2-seeded ribbed fruits

Habitat: moist shady places; common in the East Bay hills

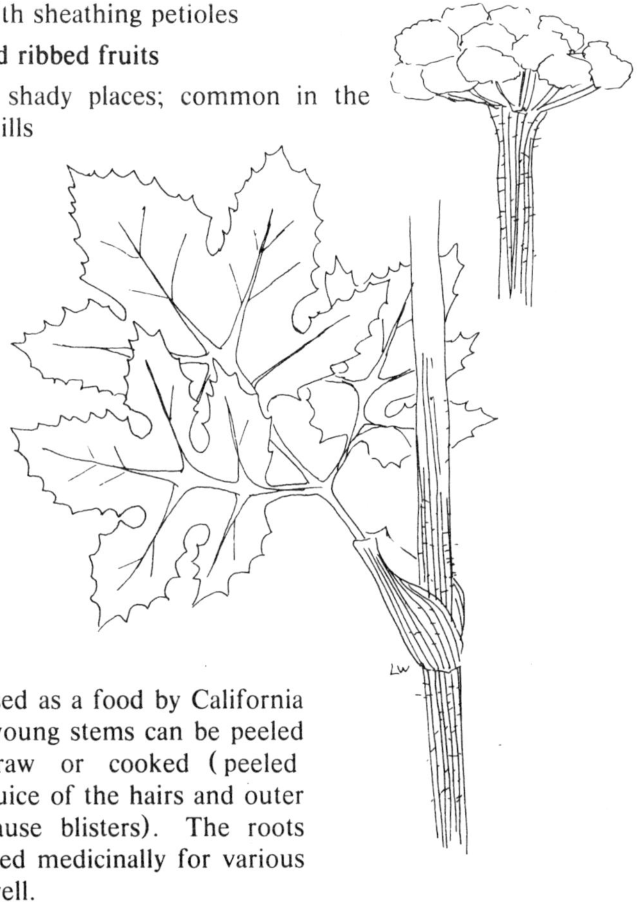

Heracleum probably refers to Hercules, reputedly the first to use this plant medicinally. *Lanatum* means woolly, for the hairy stems.

It was also used as a food by California Indians; the young stems can be peeled and eaten raw or cooked (peeled because the juice of the hairs and outer stems can cause blisters). The roots have been used medicinally for various purposes as well.

Umbelliferae (Parsley family)

Conium maculatum

Poison Hemlock

Biennial, introduced from Europe, Asia

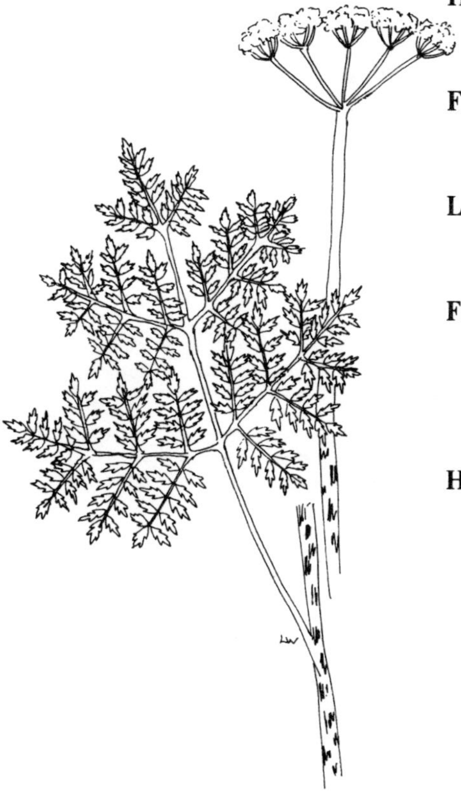

Habit: stout, branching stems 1-4 m. high, purple-streaked or spotted

Flowers: small, white, in compound umbels (April-July)

Leaves: large, finely dissected into many leaflets (tripinnate)

Fruits: schizocarps, each made up of 2 prominently ribbed mericarps (one-seeded portion of a fruit that splits at maturity)

Habitat: roadsides, waste places, fields

Coneion is the ancient Greek name of this plant, which was the source of the Athenian state poison used by Socrates. All parts of this plant are extremely poisonous because of the alkaloid coniine. Many other umbellifers contain poisonous alkaloids in varying amounts. The species is distinguishable from other large umbellifers by the purple blotches present on its stems, and the finely divided leaves. Wavy longitudinal ribs on the mericarps and a mouse-like odor also characterize this species.

Stellaria media

Chickweed

Annual, introduced from Europe

Habit: branched, spreading stems, 1-3 dm. long, with a single line of hairs on one side

Flowers: tiny, white, with 5 deeply-cleft petals which are shorter than sepals (all year)

Leaves: opposite, ovate

Fruit: capsule opening by 6 short teeth

Habitat: lawns, gardens, waste places

The genus name comes from the Latin *stella* meaning *star* for the star-shaped flowers. The common name *Chickweed* has been in use since the 15th century, and comes from the use of this plant as food for chickens, geese, and cage birds. This species is extremely common and widespread throughout the temperate zone and produces many seeds very quickly. The plant may be cooked as a vegetable or eaten raw in salads.

Caryophyllaceae (Pink family)

Marah fabaceus

California Manroot, Wild Cucumber

Perennial, native

Habit: viny; stout, trailing stems, 1-6 m. long, climbing by tendrils

Flowers: buff white, unisexual, ovary inferior, 5-petaled (April-July)

Leaves: large, entire with 5-7 pointed lobes and a U-shaped base

Fruits: globe-shaped with many surface spines

Habitat: hillsides, fields, occasional in yards

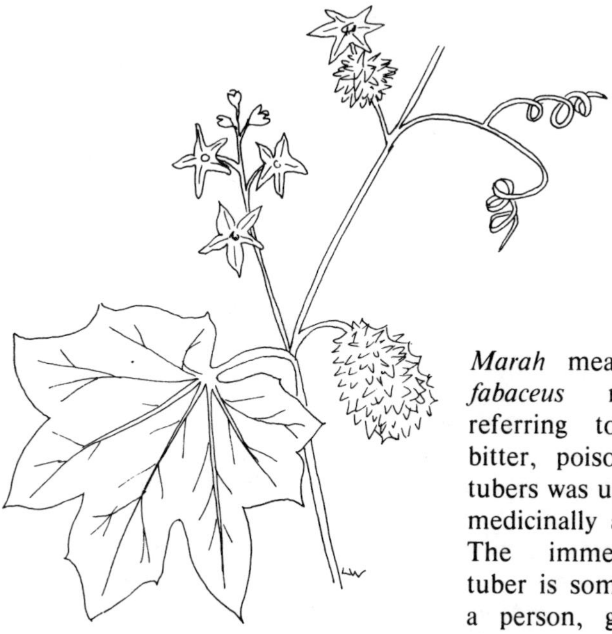

Marah means bitterness and *fabaceus* means bean-like, referring to the seeds; the bitter, poisonous juice of the tubers was used by Indians both medicinally and to stupify fish. The immense underground tuber is sometimes shaped like a person, giving this species one of its common names.

Cucurbitaceae (Cucumber family)

Montia perfoliata

Miner's Lettuce, Indian Lettuce

Native annual

Habit: erect stems, 1-3 dm. high, succulent

Flowers: white, in clusters, each with a pair of fleshy sepals (Jan.-July)

Leaves: stem leaves cup-shaped below the flower clusters; lower leaves come from a basal rosette

Fruit: capsules with 3 shiny, black seeds

Habitat: widespread in moist, shady places; woods

Montia is named after an Italian botanist, Guiseppe Monti; *perfoliata* refers to the shape of the stem leaves with the stems running through the center. The common names come from the plant's use as a salad vegetable; the fleshy leaves and stems make a good lettuce substitute. The other common *Montia* species in this area is *Montia sibirica*, which has small oval opposite leaves without petioles on the stems.

Portulacaceae (Purslane or Portulaca family)

Fragaria chiloensis

Beach Strawberry

Perennial, native

Habit: horizontal stems from stout rootstocks

Flowers: white to pinkish, 5 sepals, 5 petals (Mar.-Aug.)

Leaves: 3 leaflets, smooth and shiny above, silvery below, leathery

Fruit: juicy, red receptacle with many achenes

Habitat: commonly planted as a ground cover; also common on sand dunes, beaches

Chiloensis refers to an island off Chile, Chiloe Island, where this species also grows. The Chilean form of this species is one of the parents of the domestic strawberry, which is a hybrid. *Fragaria* comes from the Latin word meaning fragrant. *Strawberry* is derived from the Old English word *streawberige* meaning *chaffy berry*, which probably refers to the achenes that protrude from the berry.

Rosaceae (Rose family)

Rubus procerus

Blackberry,
Himalaya Berry

Perennial, introduced from Europe

Habit: sprawling, evergreen shrubs to 3 m. tall, stems rooting at nodes, very prickly

Flowers: white to pinkish, 5 petals and sepals with many stamens and pistils (May-Sept.)

Leaves: 3-5 leaflets, generally 5 on vegetative shoots and 3-5 on flowering shoots, stipulate

Fruits: black, shiny, and globular; a multiple fruit made up of succulent drupelets

Habitat: waste places, disturbed areas, eroded banks

Rubus means bramble in Latin; *bramble* comes from the Old English word meaning wiry or thorny shrub. This plant can spread rapidly over disturbed areas, forming thickets, especially on poor acidic soils. The berries are excellent to eat, and make good pies and jams.

Rosaceae (Rose family)

Convolvulus arvensis

Field Bindweed, Orchard Morning Glory

Perennial, introduced from Europe

Habit: trailing stems, twining and sometimes climbing

Flowers: white to pink, funnel-shaped on long stems; flowers open during the day and close at night

Leaves: triangular, arrowhead-shaped with pointed lobes

Fruits: capsules, with 4 large seeds per capsule

Habitat: fields, waste places, roadsides

Convolvere, meaning *to twine*, gives this genus of twining and trailing herbs its name; *arvensis* means of arable land. This species is a serious weed of cultivated areas in California and other parts of the world. Its ability to spread underground and to send up shoots from root buds makes it hard to eradicate. The roots themselves are tuberous and can become quite large; consequently they can continue producing shoots for a long time, even when the shoots are continually removed.

Convolvulaceae (Morning Glory family)

Capsella bursa-pastoris

Shepherd's Purse

Annual, naturalized from Europe

Habit: extremely variable in size and form, stems generally erect, 1-5 dm. high

Flowers: tiny, white, inconspicuous, in long racemes (all year)

Leaves: toothed lower leaves, in a rosette; leaves on stem smaller and untoothed

Fruit: heart-shaped silicles

Habitat: gardens, fields, waste places

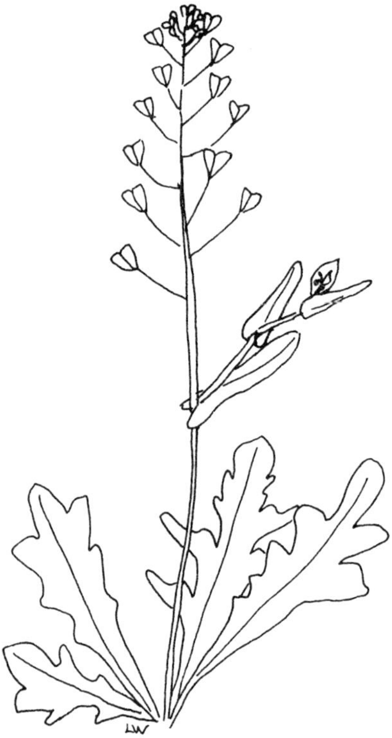

This is an extremely common species world-wide. *Capsella* is Latin for *little box*. The specific epithet, *bursa-pastoris* means shepherd's purse; the fruits resemble the bag-shaped purses that were once worn by shepherds. The flowers are self-pollinated before they open. The seeds produce a sticky mucilage when wet, which increases their dispersability. Young leaves are edible and have a peppery taste. Extracts from this plant have marked anti-inflammatory properties; a cotton ball soaked in juice from crushed plants can be used to stop nosebleeds.

Cruciferae (Mustard family)

Polygonum aviculare

Prostrate Knotweed

Annual, introduced from
Europe and Asia

Habit: branched, sprawling stems from a tough branching taproot

Flowers: inconspicuous white, pale-green, or pinkish flowers in clusters, in leaf axils (all year)

Leaves: simple, bluish-green, small, with silvery bracts at the base

Fruits: 3-angled achenes

Habitat: trampled places, gardens, waste places

In sandy soils, the tap root may penetrate up to 3 feet. The shoots are tough and wiry and resistant to treading damage; they also have an unusual capacity (for annuals) to regenerate the shoot if the top is cut off during the growing season. Both of these characteristics contribute to this plant's success in trampled areas. The fruits can remain viable for many years in the soil, and will germinate under favorable conditions.

Polygonaceae (Buckwheat family)

Picris echioides

Bristly Ox Tongue, Bugloss

Biennial, introduced from Europe

Habit: angled upright stems covered with barbed hairs

Flowers: yellow, in heads surrounded by long, densely bristly bracts (June-Dec.)

Leaves: long, narrowly oblong, rough and hairy with noticeable prickly bumps

Fruits: oblong orange-colored achenes with abundant pappus

Habitat: grassy areas, sidewalk edges, waste places

The rough tongue-like leaves give this plant its common name; *Bugloss* means ox tongue in Greek and is a name given to several plants with rough tongue-like leaves. The large, noticeably bumpy leaves distinguish this species. *Picros* means bitter.

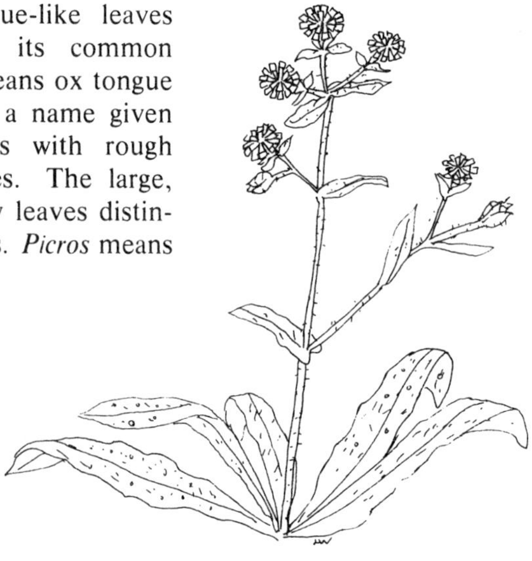

Compositae (Sunflower family)

Cotula australis

Australian Brass-Buttons

Annual, naturalized from Europe

Habit: slender, branching stems, .3-2 dm. high

Flowers: yellow, in flattened disk-shaped heads, inner rows bisexual, outer rows female

Leaves: finely dissected into linear segments, sparsely hairy

Fruits: achenes without pappus

Habitat: disturbed areas along sidewalks, in gardens, waste places

Brass-Buttons refers to the button-shaped flower heads; a related species, *Cotula coronopifolia*, is much brighter yellow and showier. The genus name *Cotula* refers to the cup-shaped depression at the base of the leaves in *C. coronopifolia*; *kotule* means *little cup* in Greek.

Compositae (Sunflower family)

Senecio vulgaris

Common Groundsel, Golden Ragwort

Annual, introduced from Europe

Habit: erect, 1-5 dm. high

Flowers: pincushion-like yellow flower heads, in clusters; each head surrounded by a row of long linear black-tipped bracts (all year)

Leaves: pinnately lobed leaves, clasping, slightly auricled

Fruits: achenes with pappus

Habitat: garden beds, waste places

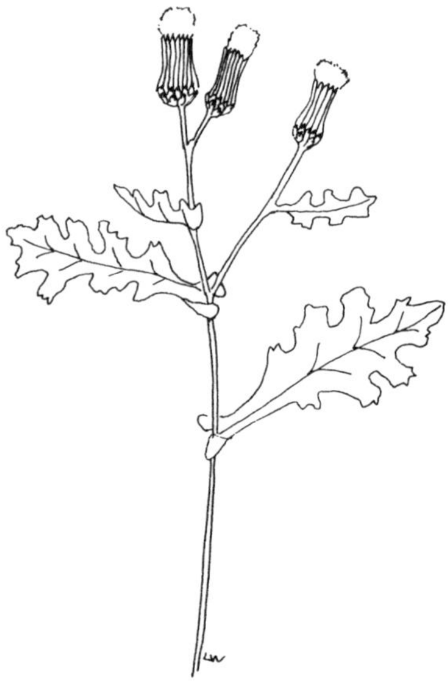

Senecio is derived from the Latin word *senex* which means *old man*; the name probably refers to the gray pubescence of some species. *Groundsel* comes from the Old English word meaning *ground swallower*, an apt name for such a common weed. Infusions of the ground herb were used to speed childbirth by both Europeans and American Indians. The alkaloids in the plant are poisonous to cattle and horses if a large amount is consumed.

Sonchus oleraceus

Common Sow Thistle

Annual, introduced from Europe

Habit: erect, smooth-stemmed, branched, 5-35 dm. high

Flowers: pale yellow, head composed only of ray flowers (all year)

Leaves: upper leaves sessile, clasping the stem; leaves are lobed, with spiny teeth

Fruits: achenes with a crown of pappus hairs

Habitat: fields, roadsides, waste places

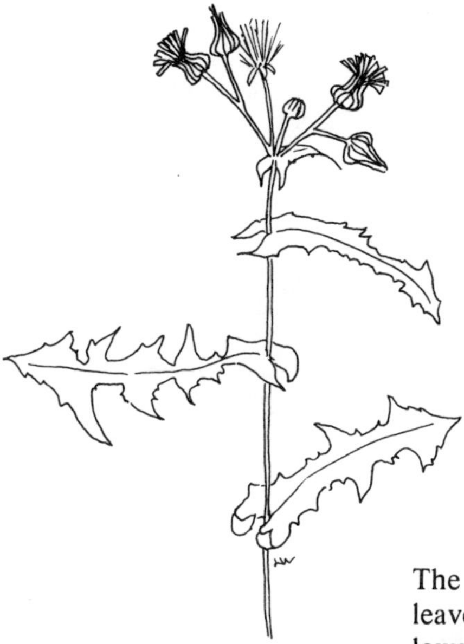

The pointed basal lobes on the upper leaves and the triangular tip on the lower leaves distinguish this species from the otherwise similar species, *S. asper.*

When bruised, plants exude a bitter, milky sap; in spite of this, rabbits, goats, and pigs will readily eat them. In Europe, young sow-thistle leaves are eaten in salads and soups. They also are mashed up and fed to ducks and geese. Sow-thistles are generally naturalized throughout inhabited areas in temperate and sub-tropical regions of the world.

Compositae (Sunflower family)

Taraxacum officinale

Dandelion

Perennial, native of Europe

Habit: leaves in a basal rosette, flowering stalks 1-4 dm. high

Flowers: bright yellow, in heads (all year)

Leaves: simple, basal, spatulate to oblanceolate, irregularly toothed

Fruits: achenes with a soft fluffy pappus borne on a long beak

Habitat: lawns, gardens, disturbed areas

The common name *dandelion* comes from the French *dent-de-lion* meaning *tooth-of-the-lion*, which refers to the jagged leaf edges. Dandelion leaves can be eaten as greens in early spring and the flowers can be used to make wine. The stout tap root has been used medicinally; the active property is a bitter compound called *taraxin*. The seeds can be formed without fertilization (apomictically), ensuring ample seed production. At the time of flowering, the scape is often quite short, elongating rapidly when the seeds are ready to be dispersed. New shoots may also regenerate from the tap root if the rosettes are destroyed. The rapid maturation of seeds (two weeks) and ability to regenerate from the roots make this species well-suited to life in lawns.

Compositae (Sunflower family)

Medicago polymorpha

Bur Clover

Annual, introduced from Europe

Habit: stems trailing from basal rosettes, 1-4 dm. long

Flowers: yellow, small, pea flower-like (Feb.-May)

Leaves: 3 leaflets with toothed margins; stipules at base of leaves with prominent teeth

Fruits: spiny, sometimes smooth, coiled pods; 3-6 seeds per pod

Habitat: fields, waste places, sidewalk edges, gardens

This species is a relative of alfalfa, *M. sativa*, and, similar to alfalfa, is also used as a forage plant because of its nutritive value. The nitrogen-fixing bacteria associated with its root nodules make the plant more nutritious and enrich the soil. This species can be distinguished from a closely related species, *M. arabica*, because of the lack of prominent dark purple blotches on the leaves, which are characteristic of *M. arabica*.

Leguminosae (Pea family)

Brassica campestris

Field Mustard, Yellow Mustard

Annual, naturalized from Europe

Habit: erect stems, 3-12 dm. high

Flowers: pale yellow, 4 petals; flowers in loose clusters (Jan.-June)

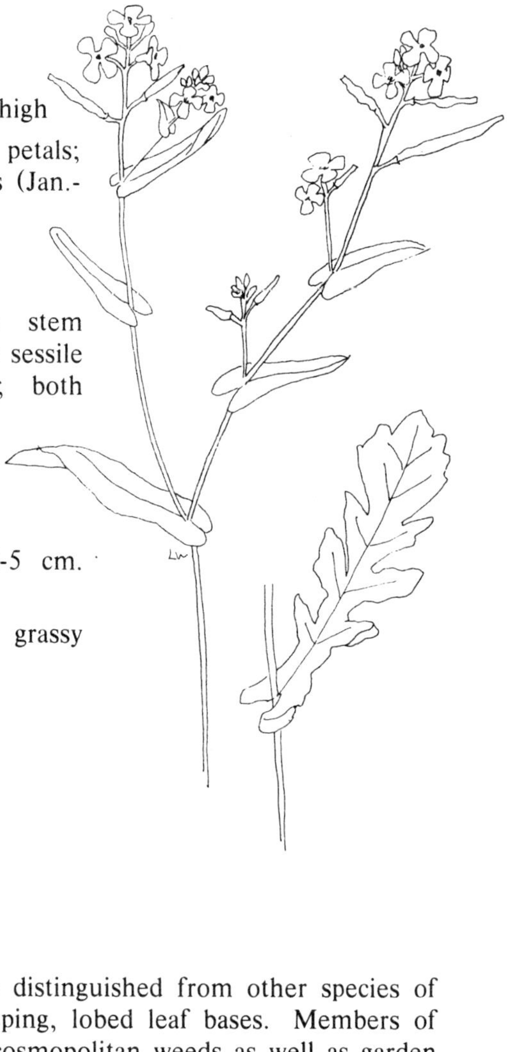

Leaves: lower leaves on stem pinnatifid, upper leaves sessile and arrowhead-shaped; both with clasping bases

Fruits: rounded siliques, 2-5 cm. long, with a stout beak

Habitat: fields, roadsides, grassy areas, waste places

This species can be distinguished from other species of *Brassica* by the clasping, lobed leaf bases. Members of this genus include cosmopolitan weeds as well as garden vegetables (varieties of *Brassica oleracea* include broccoli, cauliflower, cabbage, collards, and Brussels sprouts). *Campestris* means *of fields.*

Cruciferae (Mustard family)

Camissonia ovata

Sun-cups, Golden Eggs

Perennial, native

Habit: leaves in upright basal rosette, corolla tube 5-10 cm. long

Flowers: bright yellow, 4 petals forming a cup at the top of a long floral tube (March-June)

Leaves: ovate, with long petioles, leaves 6-20 cm. long, occasionally with reddish main veins

Fruits: capsules, with few seeds

Habitat: open grassy slopes, fields

The ovaries of this species are hidden in the basal rosette of leaves, sometimes as many as 6 cm. below the ground; how seeds are dispersed is a bit of a puzzle. This plant blooms in the day unlike most other members of this family.

Onagraceae (Evening Primrose family)

Centaurea solstitialis

Yellow Star Thistle, Barnaby's Thistle

Annual, introduced from Europe

Habit: rigid, spreading stems branched from the base, 3-7 dm. high; stems appear whitish and woolly

Flowers: small, bright yellow, in solitary heads, surrounded by long rigid spines, 3-pronged on lowest bracts (May-Oct.)

Leaves: basal leaves deeply lobed; lower leaves long, entire, sharply pointed

Fruits: achenes of two types produced: light-colored with pappus, and dark without pappus

Habitat: roadsides, vacant lots, fields

Centaurea is derived from the Greek word meaning *herb of the centaurion*; a relative of this species was said to have cured the centaur Chiron, a mythological healer, after an arrow belonging to Hercules fell on his foot. The second common name comes from the flowering time, which occurs in Europe around July 11, St. Barnaby's Day. The smaller dark achenes are hard to separate from commercially used seeds; consequently, they are easily introduced to new areas as an impurity.

Compositae (Sunflower family)

Oxalis corniculata

Creeping Wood Sorrel

Perennial, introduced from Europe

Habit: creeping, prostrate stems, .5-3 dm long, rooting at nodes

Flowers: small, yellow-orange, funnel-shaped (all year)

Leaves: 3 leaflets, hairy on the underside, occasionally purplish

Fruits: cylindrical capsule, 1-2.5 cm. long

Habitat: common in lawns, along sidewalks, in gardens, especially in shady places

The genus name comes from *oxus*, meaning sour; the foliage of *Oxalis* species tastes sour due to high concentrations of oxalic acid. The seeds are ejected from the fruits by a build-up of tension within the seed pod; seeds may be catapulted up to 1 m. away. The leaflets of *Oxalis* species fold up at night and open during the day.

Oxalidaceae (Wood Sorrel family)

Oxalis pes-caprae

Bermuda Buttercup, Wood Sorrel

Perennial, introduced from South Africa

Habit: leaves in rosettes, leafless flowering stalks, 1.5-3 dm. high

Flowers: lemon-yellow, funnel-shaped, borne in umbels (Nov.-May)

Leaves: 3 leaflets, clover-like with notch in each leaflet, often with red blotches or spots

Habitat: gardens, flower beds, lawns

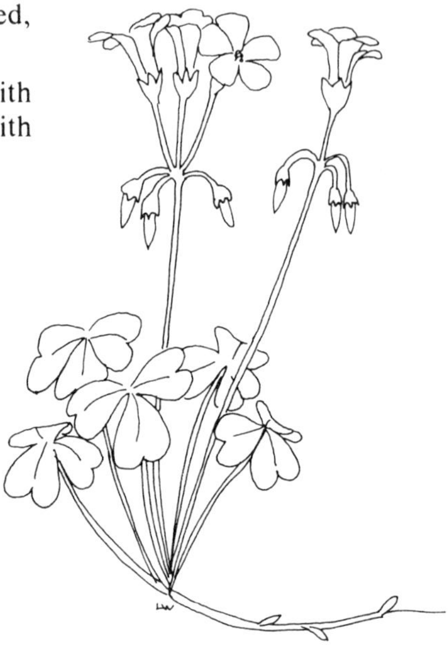

Oxalis means sour in Greek; the leaves and stems of *Oxalis* species have a refreshing sour taste due to high concentrations of oxalic acid. The abundance of this species is due in part to its vigorous vegetative spread by means of bulbils produced on underground rootstocks. The bulbils fall off and will persist in the soil, sending up new shoots in the fall. In California, this species reproduces exclusively by bulbils.

This species has a complex breeding system with 3 flower forms, one of which predominates here. Crossing between the forms is necessary for optimal seed production. Therefore, few seeds are produced in this area in spite of copious flowering. The bulbils are sometimes used as a vegetable in southern France and northern Africa. Leaflets of *Oxalis* species exhibit *sleep movements*- the leaflets fold downward at night and in cold weather.

Oxalidaceae (Wood Sorrel family)

Duchesnea indica

Mock Strawberry

Perennial, ornamental

Habit: stems creeping, rooting at nodes

Flowers: yellow, 5 petals (Mar.-Aug.)

Leaves: alternate, trifoliate, dark green, glossy

Fruits: spongy, non-juicy strawberry-like receptacle with numerous achenes on surface

Habitat: ornamental, in gardens

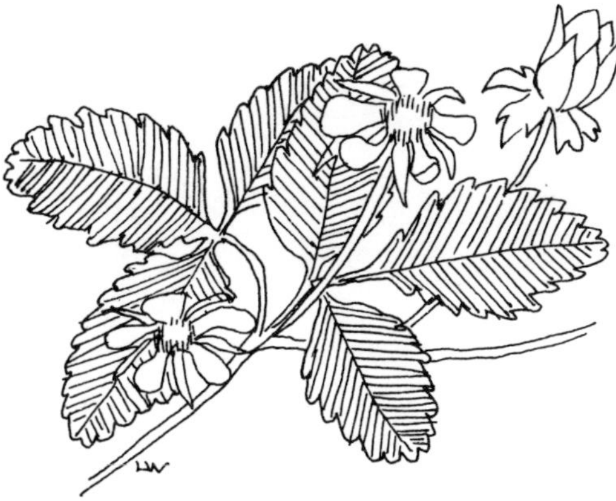

This species is named after A.N. Duchesne, the monographer of the genus *Fragaria*, to which the true strawberry belongs. The leaves of this species are similar to those of *Fragaria*, but look for the absence of silvery hairs on the underside of the leaves, presence of yellow rather than white flowers, and non-juicy fruits.

Rosaceae (Rose family)

Ranunculus californicus
California Buttercup

Perennial, native

Habit: erect, well branched stems, 2-6 dm. high

Flowers: yellow, shiny, with 9-16 (or more) petals and many stamens (Feb.-May)

Leaves: 3-5 lobed basal leaves with long petioles, variable

Fruits: achenes, with a stout curved beak

Habitat: moist grassy areas, meadows, fields

The genus name comes from the Latin word meaning *little frog*, due to the moist habitat typical of the genus. This species varies in appearance (height, number of petals, leaf shape) from coastal to inland plants. The numerous shiny petals are characteristic of the species. *R. repens*, an introduced European species, can be distinguished from this species by its creeping habit and 5-petaled flowers.

Ranunculaceae (Buttercup family)

Eschscholzia californica

California Poppy

Annual to perennial, native

Habit: stems spreading, 2-6 dm. long

Flowers: yellow-orange, 4 petals, cup-shaped; sepals are united into a single calyx that is quickly lost as the corolla expands (Feb.-Sept.)

Leaves: finely dissected, gray-green

Fruits: capsules, 3-8 cm. long

Habitat: fields, open areas, along roadsides

This species is named for a German botanist, J.F. von Eschscholtz, a doctor who traveled with the 1815 Russian expedition to California. *E. californica* is the state flower of California and has become increasingly abundant, because moderate levels of disturbance favor its spread. The species has also become naturalized in other countries after being introduced as a garden plant. This plant was used by California Indians as a pain-killer, because of its narcotic properties.

Papaveraceae (Poppy family)

Tropaeolum majus

Nasturtium

Perennial, cultivated, from Central and South America

Habit: prostrate stems, climbing by means of twining petioles

Flowers: yellow-orange, showy, irregular, borne in leaf axils

Leaves: alternate, peltate

Fruit: 3-seeded schizocarp

Habitat: cultivated in gardens

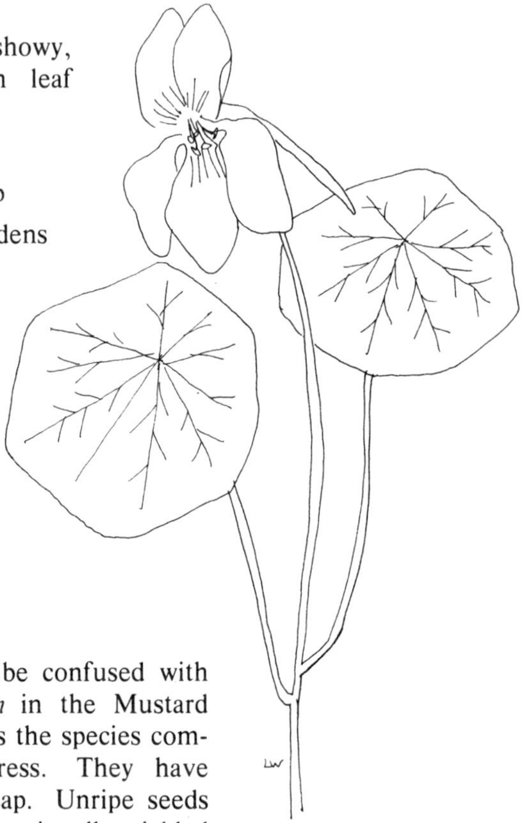

This plant shouldn't be confused with the genus *Nasturtium* in the Mustard family, which includes the species commonly called watercress. They have similar oils in their sap. Unripe seeds of *Tropaeolum* are occasionally pickled and used like capers. The name *nasturtium* is derived from the Latin words meaning *nose-twister* or *nose-torturer.*

Tropaeolaceae (Nasturtium family)

Cirsium vulgare

Bull Thistle, Common Thistle

Biennial, introduced from Europe

Habit: erect, stout, branching stems, 6-12 dm. tall

Flowers: red-purple, in heads surrounded by lanceolate bracts; heads in terminal clusters, 4-5 heads per cluster (June-Oct.)

Leaves: spiny leaf blades, coarsely toothed or deeply lobed to midrib, green on upper surface, lighter and somewhat hairy on lower surface

Fruits: achenes with plumose pappus

Habitat: waste places, fields

Thistle is derived from an Indo-Eủropean base word *thista* meaning *to prick.* Plants first appear as a broad rosette of spiny leaves in the fall, and then overwinter accumulating food reserves in a deep tap root. Flowering occurs the following season. The fruiting head of achenes is similar to that of a dandelion, but is larger and denser. Thistles are an emblem of Scotland; legend has it that they were instrumental in repelling Norse invaders.

Compositae (Sunflower family)

Polygonum persicaria

Lady's Thumb, Heartweed

Annual, introduced from Europe

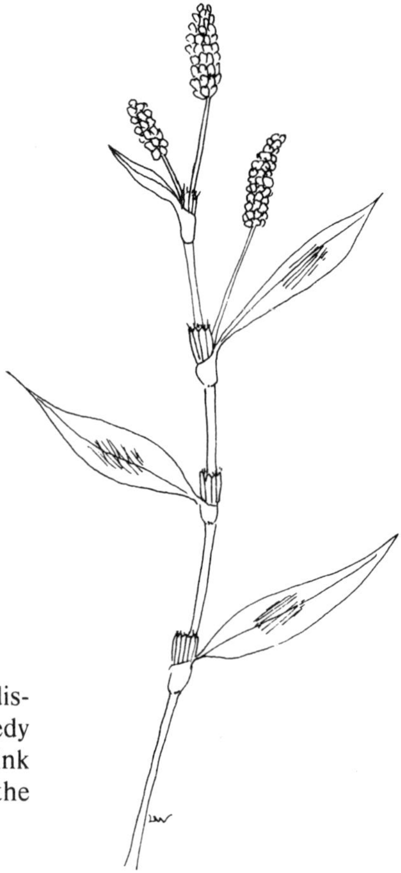

Habit: erect or sprawling, 2-8 dm. high

Flowers: dusty pink to dark purple, tiny tubular flowers, in dense clusters, one to several spikes per plant (all year)

Leaves: simple, usually with dark splotches in the center; fringed papery sheath at nodes (ocrea)

Fruits: achenes, 3-angled, black and shining

Habitat: garden edges, moist waste places

This species can be distinguished from other weedy *Polygonum* species by its pink flowers and blotches on the leaves.

The characteristic swollen leaf nodes give this species its generic name - from the Greek words meaning *many knees.* The common name of knotweed also refers to the swollen appearance of the nodes. The other common name comes from the somewhat thumb-like appearance of the flower spikes. This species grows throughout the U.S. in moist waste places.

Polygonaceae (Buckwheat family)

Centranthus ruber

Red Valerian

Perennial, introduced from
S. Europe and Asia

Habit: compact and bushy stems, 3-8 dm. high, smooth-stemmed

Flowers: small, red, pink, or white; with tubular corolla, inferior ovary, 1 stamen, in dense terminal clusters

Leaves: opposite, ovate, bluish-green

Fruits: hard achene-like nuts with attached calyx bristles that promote wind dispersal of the seeds

Habitat: gardens, yards, occasional escape

The common name is derived from the medieval Latin meaning *plant from Valeria*, a Roman province in western Hungary. Red, pink, and white-flowered forms are found in our area; hummingbirds visit only red and pink flowers, whereas moths visit white flowers preferentially. Butterflies visit all three, in different frequencies.

Valerianaceae (Valerian family)

Erodium cicutarium

Red-stem Filaree, Cranesbill

Annual, introduced from the Mediterranean region

Habit: reddish stems and leaves in basal rosettes, 1-4 dm. high

Flowers: rose-purple, 5-merous (Feb.-May)

Leaves: leaflets pinnately dissected (leaves twice pinnate) and toothed

Fruits: schizocarps made up of 5 mericarps attached to a long, elastically dehiscent beak

Habitat: widespread, in open areas, grasslands, roadsides

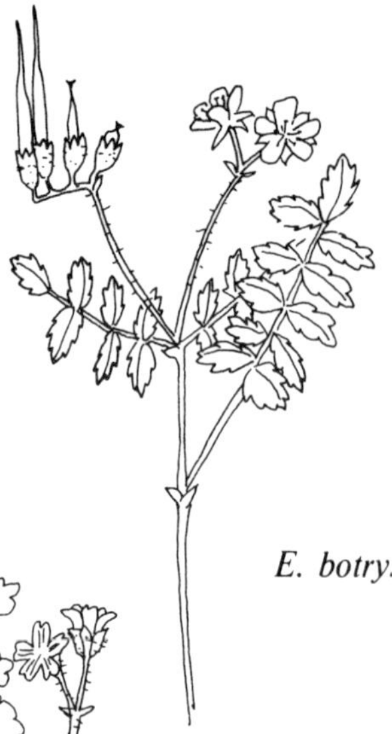

E. botrys

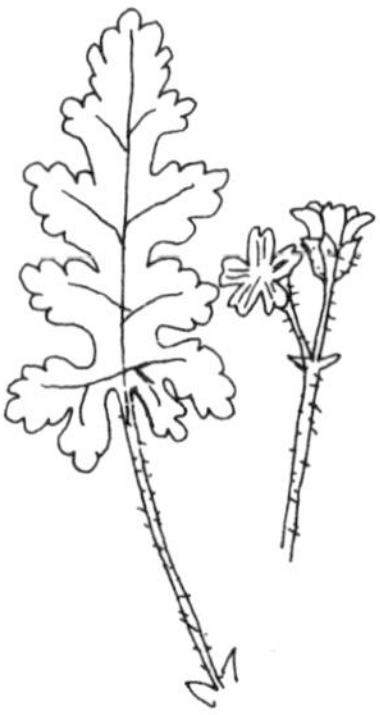

E. moschatum

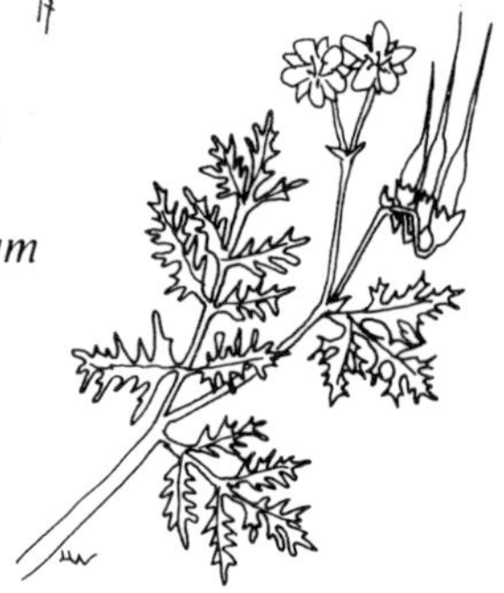

E. cicutarium

E. cicutarium is an important forage plant in California annual grasslands. It was one of the first alien species to be introduced by the Spanish. Plant size varies depending on the favorableness of the site. Foliage has been used in warm herbal baths to treat pains of rheumatism and also to make a diuretic tea. When the beak dries out, the 5 barbed mericarps split apart and their long bearded styles become spirally twisted; they then can easily stick to the fleece and fur of animals (and to socks!) *E. moschatum*, White-stem Filaree, and *E. botrys*, Broad-leaf Filaree, are commonly occurring annuals also illustrated here.

Geraniaceae (Geranium family)

Vicia sativa

Spring Vetch

Annual, introduced from Europe

Habit: twining, vine-like stems, 3-8 dm. high, clinging by tendrils

Flowers: rose-purple, in pairs or borne singly (Apr.-July)

Leaves: pinnately compound, with 2-8 pairs of leaflets; each leaf with a pair of stipules at the base

Fruits: flattened pods with 2 sides, 4-6 cm. long, with 2 or more seeds

Habitat: waste places, grassy areas, garden beds

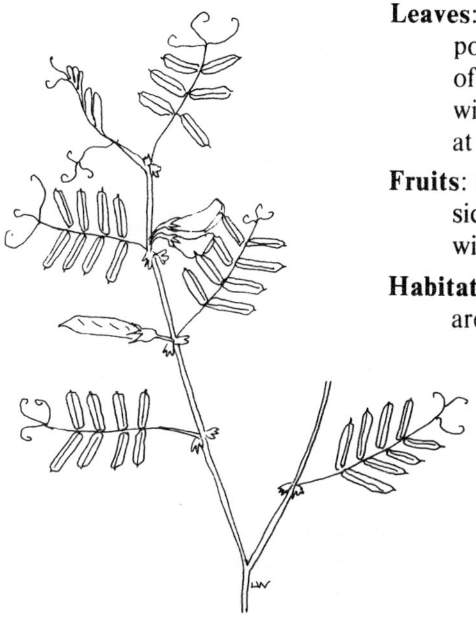

The stipules bear small, dark glands that produce nectar (called extra-floral nectaries) that are attractive to ants. When ants are present on *Vicia* plants, they tend to protect the plants from damage by herbivores. There are two common varieties of this species, var. *angustifolia* and var. *sativa.* Var. *angustifolia* has smooth foliage, narrower leaflets, and fewer pairs of leaflets. Var. *sativa* is slightly hairy, and has broader leaflets.

Leguminosae (Pea family)

Malva nicaeensis

Bull Mallow, Cheeseweed

Annual, introduced from Europe

Habit: semi-erect stems, 2-6 dm. high

Flowers: pink, petals twice as long as sepals, borne in leaf axils (Mar.-Sept.)

Leaves: heart-shaped at base, bluntly lobed with 5-7 lobes, rounded

Fruits: seed pods are flattened rings of one-seeded fleshy segments, separating at maturity

Habitat: sidewalk edges, waste places

There are 2 other common *Malva* species in this area, *M. neglecta* and *M. parviflora. Malva neglecta* is very similar to *M. nicaeensis*, but its leaves are rounder and less lobed. The stems of *M. parviflora* are more erect than those of *M. nicaeensis* and the petals are only slightly longer than the calyx. Mallow species share the common name of *cheeseweed* because of the supposed resemblance of the fruits to round cheeses. Mallow fruits and young shoots are edible, and may be used as greens or in salads. These weedy mallows are often infected with a rust fungus and may be a source of infection for cultivated mallows.

Malvaceae (Mallow or Hollyhock family)

Modiola caroliniana

Wheel Mallow, Bristly Mallow

Perennial, from the south-eastern U.S.

Habit: low, spreading stems, 3-6 dm. long

Flowers: small, brick-red, surrounded by 2-3 narrow bractlets, flowers solitary in leaf axils

Leaves: rounded, palmately lobed, coarsely toothed

Fruits: disk-shaped, made up of 15-30 bristly 2-seeded segments

Habitat: lawns, waste places

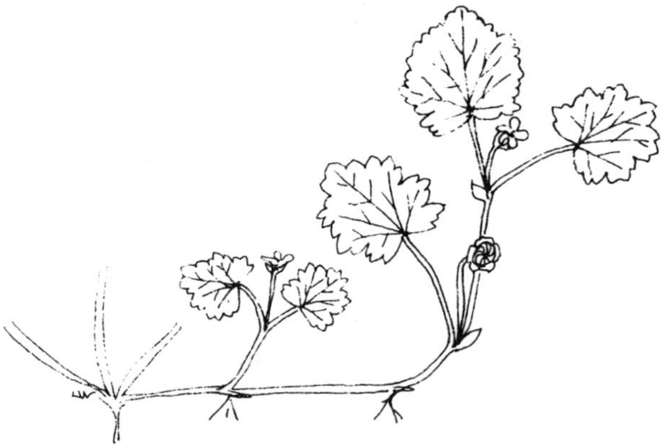

Members of the Malvaceae can be identified by their tubular column of stamens. *Modiola* can be distinguished from *Malva* species by the round terminal stigma. The lobed leaves and creeping habit also help to identify this species.

Anagallis arvensis

Scarlet Pimpernel, Poor Man's Weatherglass

Annual, introduced from Europe

Habit: sprawling, well-branched stems, .5-5 dm. in length, 4-angled in cross-section

Flowers: salmon-colored or brick-red with purple centers, in leaf axils; flowers with 5 stamens opposite petals (Jan.-June)

Leaves: opposite or in whorls of 3, simple, ovate, or deltoid

Fruits: smooth, round capsules with many seeds; top of capsule comes off like a lid (circumscissile)

Habitat: common along sidewalks, edges of gardens, in waste places

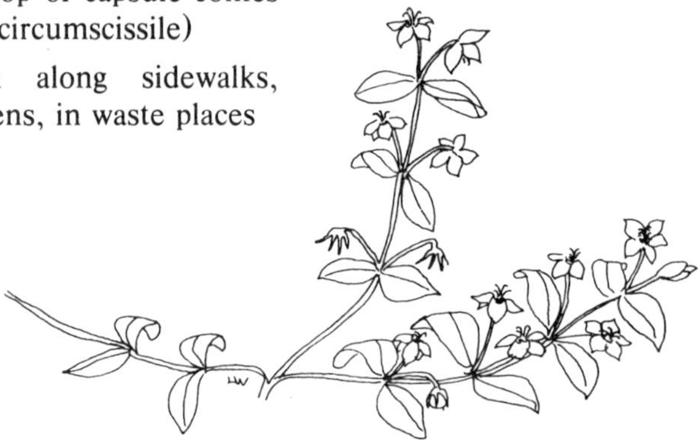

The genus name comes from the Greek words meaning *to delight in* . The flowers open only on sunny days, thus giving the species one of its common names. Occasional blue flowered individuals may be found in some populations. In Britain, the flowers are brilliantly red, thus the name *Scarlet Pimpernel*, which doesn't seem quite appropriate for our salmon-colored flowers!

Primulaceae (Primrose family)

Geranium robertianum

Herb Robert

Annual to biennial, native of Europe and Asia

Habit: ascending to upright stems, from basal rosettes, 1-3.5 dm. high

Flowers: dark pink to purple, 5-merous, borne in axils of leaves (Mar.-Nov.)

Leaves: palmately cleft in 3-5 divisions

Fruits: made up of 5 mericarps borne at the base of an elongated beak; when the beak dries out, the seeds are flung forcibly away from the plant

Habitat: moist shady areas, gardens

The stems and leaves of this species tend to be reddish and have a minty, spicy smell. The stems and leaves are sparsely hairy. The combination of forcible discharge of seeds and the tackiness of the fruits make this species a good colonizer, especially of rocky crevices in cliffs. After being introduced to the U.C. Botanical Garden around 1940, this species has spread down Strawberry Creek and into the East Bay hills, where it flourishes in moist areas.

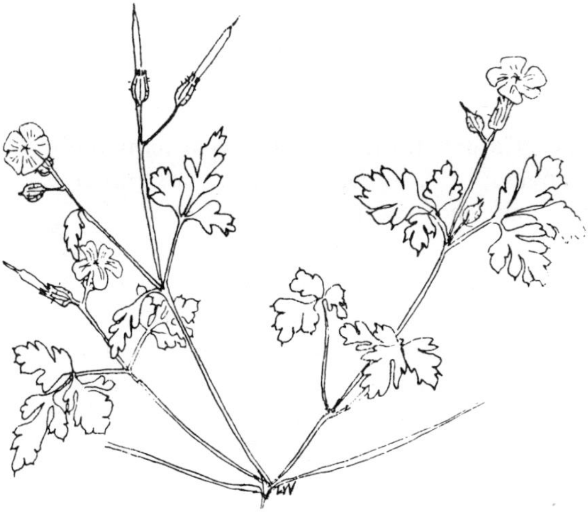

Geraniaceae (Geranium family)

Vinca major

Greater Periwinkle

Perennial, introduced from Europe

Habit: trailing, evergreen, viny

Flowers: dark blue, tubular, solitary in leaf axils

Leaves: ovate, dark green, opposite

Fruits: follicles with many seeds; plants rarely produce fruits in our area, however

Habitat: commonly used as a ground cover

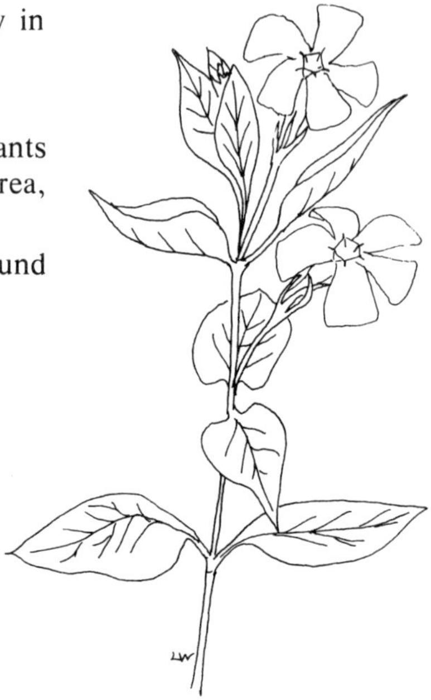

Vinca species are commonly used as ornamental ground covers. *V. major* spreads extremely rapidly vegetatively, whereas a related species, *V. minor*, which has smaller flowers, is slightly less aggressive. *V. major* is an occasional escape along streams in protected places. Both common *Vinca* species have been used externally to stop hemorrhages; this is effective because of the astringent tannins in the leaves. A powerful anti-cancer drug, vincristine, is derived from *Vinca*.

Apocynaceae (Dogbane family)

Myosotis latifolia

Common Forget-me-not

Perennial, introduced from Europe

Habit: erect hairy stems 2-5 dm. long, from underground rootstocks

Flowers: blue, wheel-shaped, in loose cymose racemes; flowers with prominent white teeth in throat of corolla (Feb.-Aug.)

Leaves: lanceolate, lower leaves with petioles, upper leaves sessile on stem

Fruits: 4 nutlets in a persistent calyx

Habitat: moist areas, gardens

Myosotis comes from the Greek words meaning *mouse-ear*, since the leaves of some species look particularly velvety. The name *Forget-me-not* comes from a 16th century German custom of wearing a blue flower to retain a lover's affection; the name became popular in English usage in the 19th century, after Coleridge used it in a popular poem. This species was initially mis-identified in our area as *M. sylvatica.* The persistent calyx that contains the nutlets is covered with slightly hooked hairs; these tend to adhere to clothing and fur, aiding in seed dispersal.

Boraginaceae (Forget-me-not family)

Veronica serpyllifolia

Thyme-leaf Speedwell

Perennial, introduced from Europe

Habit: branched, creeping stems up to 1.5 dm. high

Flowers: pale blue with 4-lobed corolla 6-7 mm. broad, in terminal racemes (Apr.-July)

Leaves: ovate to oblong, with slightly scalloped margins

Fruits: flattened capsules

Habitat: mostly in lawns

The common name *speedwell* refers to the medicinal uses of *Veronica* species. *V. officinalis* was reputed to be a strengthening and wound-healing plant and to be useful in treating coughs. The flowers and foliage of some *Veronica* species can be used in salads. There are several other *Veronica* species besides *V. serpyllifolia* that are fairly common in our area. One of these which is also found in lawns is *V. arvensis*, an annual distinguishable by its erect or ascending stems, lanceolate leaves, and bright blue flowers about half the size of the flowers of *V. serpyllifolia.*

Sherardia arvensis

Blue Field Madder

Annual, introduced from Europe

Habit: prostrate 4-angled stems, forming a mat-like rosette

Flowers: pink to blue, slender funnelform corolla with 4 lobes in clusters surrounded by closed bracts (Jan.-July)

Leaves: simple, lanceolate, in whorls of 4-6 leaves

Fruits: in 2 indehiscent one-seeded parts

Habitat: lawns, meadows, and waste places

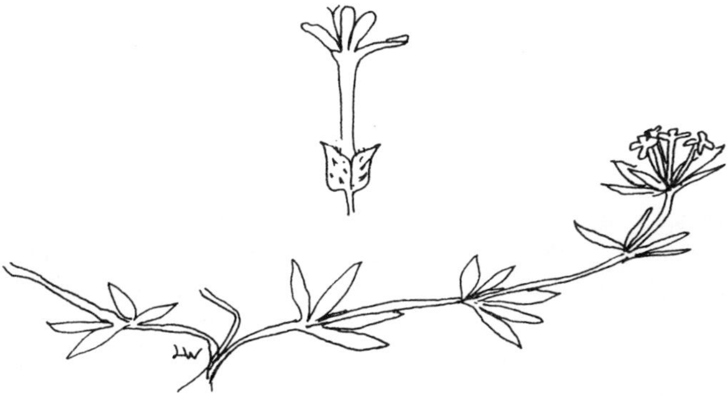

This is a fairly inconspicuous plant; look for it in lawns, where it is often quite abundant.

Rubiaceae (Madder family)

Cymbalaria muralis

Ivy-leaved toadflax

Perennial, introduced from southern Europe

Habit: trailing herb with long-petioled leaves and flowers

Flowers: purple to white, in axils of leaves; flowers have a short spur and a prominent upper lip; 4 stamens of 2 lengths (Mar.-Sept.)

Leaves: ivy-shaped, palmately lobed and veined

Fruits: capsule opening by two lateral pores

Habitat: cultivated; often on walls and banks; occasional escape in moist cool places

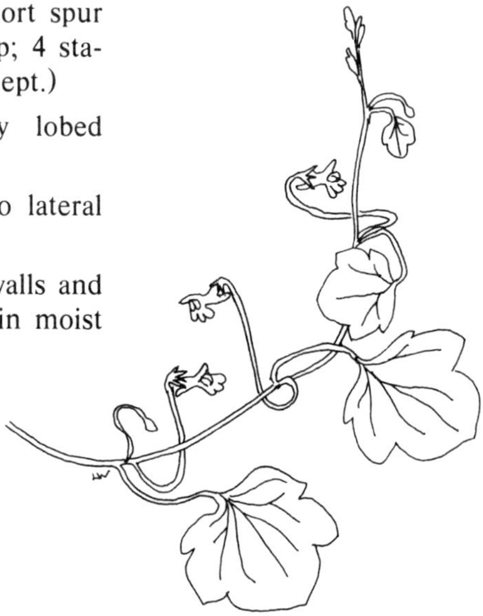

The specific epithet *muralis* refers to the wall habitat of this species. *Toadflax* is a translation of the German name meaning *wild, useless flax* or a *flax for toads.* Introduced into Britain from S. Europe as a garden plant in the early 1600's, this species subsequently has become widespread on old walls throughout Britain. The plant was brought to this country as an ornamental and is a local escape in areas of mild climate. The flower-bearing pedicels bend towards the light; after the fruits develop, the stalks bend away from the light and tend to bend downwards, so that seeds, when they are released, often fall into cracks and crevices, where they germinate and grow. Fragments of plants also readily take root and establish new plants.

Scrophulariaceae (Figwort family)

Sisyrinchium bellum

Blue-eyed Grass

Perennial, native

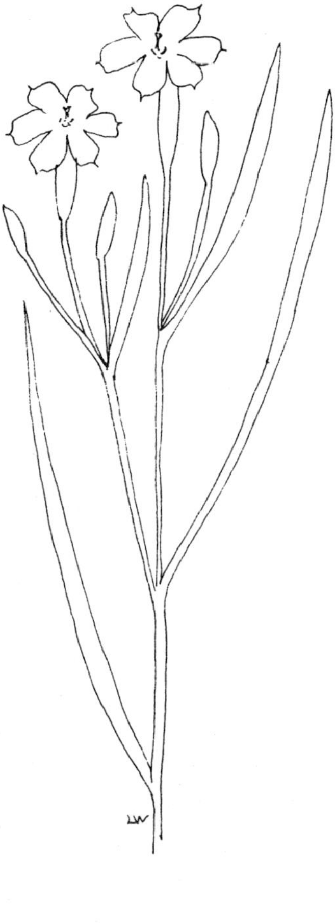

Habit: erect, branched stems, 1-4 dm. high, slender, somewhat compressed

Flowers: 6-petaled, dark purple with yellow center, filaments of stamens united (Feb.-July)

Leaves: grass-like, narrow

Fruit: capsule, 2-7 mm. in length

Habitat: open grassy areas

Sisyrinchium is an ancient name given to a plant related to *Iris* by Theophrastus; *bellus* means pretty. Flowers open during the day and close at night.

Iridaceae (Iris family)

Hedera helix

English Ivy

Perennial, introduced from Europe and Asia

Habit: vine, climbing on trees, walls

Flowers: pale green, in umbels

Leaves: vegetative leaves 3-5 lobed; leaves on flowering shoots unlobed; leaves leathery and shiny

Fruits: smooth, globose, purplish black

Habitat: planted as a ground cover

There are many ornamental cultivars of this common species, thought to be native to the dark understory of deciduous forests in Europe and Asia. Plants are able to tolerate extremely low light conditions, although the upright flowering shoots are only produced under high light conditions. The resinous sap has been used medicinally to heal sores and to relieve toothaches. Crushed leaves can be used in compresses. The young stems are a good source of yellow and brown dyes. The flowering (adult) phase is illustrated on the left; the climbing (juvenile) phase is on the right.

Araliaceae (Ginseng family)

Poa annua

Annual Bluegrass

Annual, introduced from Europe

Habit: tufted, with light green flattened stems, .5-3 dm. tall

Flowers: 3-6 per spikelet, arranged in a loose inflorescence called a panicle (Jan.-July)

Leaves: smooth, with a blunt apex

Fruit: single grain (caryopsis)

Habitat: open ground, waste places, lawns, pastures

The common name *grass* comes from a group of Old English words meaning *all green stuff eaten by cattle*; *Poa* is a Greek word meaning grass or fodder. *Poa annua* is a widely distributed grass of low elevation temperate areas. Another widespread *Poa* species is *P. pratensis*, Kentucky Bluegrass. The blunt-tipped leaves are characteristic of this genus. *Poa annua* is capable of flowering and fruiting all year in our area if moisture is available; thus, several generations may be produced within a year.

Gramineae (Grass family)

Plantago lanceolata

Narrow-leaved Plantain

Perennial (to annual), from Europe

Habit: erect flowering stems, leaves in basal rosette, 1-8 dm. high

Flowers: small, brownish white, with bracts, in dense spikes (all year)

Leaves: lanceolate, with parallel veins making the ribs heavily veined

Fruits: capsule with two seeds in chaffy appearing spikes

Habitat: open areas, empty lots, roadsides

The size of these plants varies a lot depending on the favorableness of the site. The lower flowers in a spike bloom first and then are replaced by the brownish chaffy capsules. The position of the conspicuously exserted stamens relative to the spike tells you how far flowering has proceeded. Some plants may bear only female flowers; these can be recognized by the absence of stamens. This species is also variable in its life cycle, tending to be perennial towards the coast and annual inland. This species can be distinguished from *Plantago major* by its narrower upright leaves.

Plantaginaceae (Plantain family)

Plantago major

Broad-leaved Plantain, Waybread

Perennial, introduced from Europe

Habit: leaves in prostrate rosettes, flowering stalks

Flowers: small, in dense spikes

Leaves: ovate to oval

Fruits: capsule with 6-18 seeds in brownish spikes

Habitat: lawns, disturbed places

Plantains have a long history as travelers; the name *Plantago* is derived from a Latin name meaning *footprint*. *Plantago* species have been used medicinally since antiquity, the leaves as a dressing for wounds and the seeds as a laxative because of their ability to absorb water and produce mucilage. The mucilage is easily observed if a seed is put in a drop of India ink; the exuding mucilage forms a halo in the ink around the seed. This mucilage also helps seeds adhere to the feet of birds and other passersby, thereby also dispersing the seeds.

Plantaginaceae (Plantain family)

Rumex acetosella

Sheep Sorrel

Perennial, introduced from Europe and Asia

Habit: stems erect, 1-3 dm. high

Flowers: tiny, in spikes, male flowers greenish yellow, female flowers reddish (Feb.-Aug.)

Leaves: arrowhead shaped

Fruits: small, reddish achenes with three persistent sepals, in spikes

Habitat: sandy old fields and grasslands, especially in acidic or nitrogen-deficient soil

The arrowhead-shaped leaves and rusty colored seeds distinguish this species from its close relative, *R. crispus.* The name *sorrel* comes from the Old French meaing *little acid plant*; the foliage is sour and slightly bitter because of the high concentration of oxalic acid. The young leaves can be cooked and eaten as greens or raw in salads. This species is unusual among weedy species in being dioecious (bearing male and female flowers on separate plants). Plants spread both by seeds and creeping underground rootstocks. This species is a good indicator of acidic soils.

Polygonaceae (Buckwheat family)

Rumex crispus

Curly Dock, Yellow Dock

Perennial, introduced from Europe

Habit: erect, 6-12 dm. high

Flowers: small, greenish, in dense spikes (all year)

Leaves: lower leaves lanceolate, with wavy margins, 1-2 dm. long

Fruits: red-brown, seeds surrounded by heart-shaped sepals, in dense spikes

Habitat: roadsides, waste places

The thick yellow roots of this species (and others in the genus) have been used medicinally as a remedy for anemia because of their reputed ability to take up iron present in the soil. The roots are also rich in tannins which gives them an astringent effect when applied to the skin. This species appeared in the Eastern states shortly after the English settlers arrived and was in California by 1860. Young leaves produced in the first year are sometimes cooked and eaten like spinach.

Polygonaceae (Buckwheat family)

Chenopodium album

Lamb's Quarters, Pigweed, Fat Hen

Annual, introduced from Europe and Asia

Habit: erect, branched, 2-20 dm. high

Flowers: tiny, greenish, in irregular spikes (all year)

Leaves: triangular, with whitish mealy undersides, reddish veins

Fruits: seeds of two types, one type small, black, and shiny, the other larger and brown

Habitat: gardens, fields, waste ground, sidewalk edges

This is a very variable species with numerous subspecies. The genus name is derived from the Greek word meaning *goosefoot*, which refers to the shape of the leaves. *C. album* has been used as a food plant since ancient times; Europeans valued it as a leafy vegetable until spinach was introduced from Asia in the 16th century. The seeds can be ground into a flour which resembles buckwheat flour. The seeds, especially the small black type, can be extremely long-lived and may easily persist in the soil for 500 years or more. The first two common names come from the use of this plant to feed sheep and pigs; the third common name *Fat Hen* is shared with a related species, *C. bonus-henricus* (also called Good King Henry) and refers to a greasy vegetable dish eaten in the 18th century.

Chenopodiaceae (Goosefoot family)

Matricaria matricariodes

Pineapple Weed

Annual, native

Habit: erect, branched, 1-3 dm. high

Flowers: small, greenish-yellow, in dense cone-shaped heads (Apr.-Aug.)

Leaves: lacy, pinnately divided

Fruits: achenes

Habitat: trampled sidewalk edges, gardens, roadsides

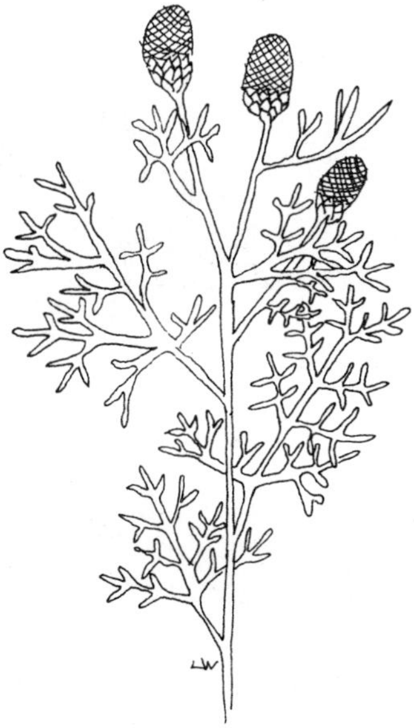

The common name comes from the strong pineapple-like odor of the foliage and from the pineapple-shaped flower heads. This is one of the few weeds native to California that has become successful along the east coast and in Europe.

Compositae (Sunflower family)

Soliva sessilis
Soliva

Annual, introduced from
South America, Chile

Habit: small, flattened rosettes of hairy branched stems,

Flowers: small, greenish, in sessile heads in axils of stems; surrounded by 7-8 hairy bracts

Leaves: pinnately dissected, feathery

Fruits: winged achenes with a pointed tip (the persistent style)

Habitat: open places, trampled areas

Soliva is the last name of a Spanish botanist. The specific epithet refers to the sessile flower heads. Within a head, the outer rows of flowers are female and lack petals; the innermost flowers are bisexual, but are sterile. This species resembles other common treading weeds with feathery dissected leaves - look for the sessile, disk-shaped flower heads in the axils of the stems.

Compositae (Sunflower family)

Coronopus didymus

Wart Cress, Swine Cress

Annual to biennial, introduced from S. America and Europe

Habit: branching stems 1-2 dm. long, in basal rosettes

Flowers: very small, greenish-white, usually with 2 stamens (Feb.-July)

Leaves: alternate, pinnately divided

Fruits: silicles conspicuously wrinkled and constricted through the center into 2 one-seeded valves

Habitat: common near sidewalks and curbsides in trampled areas, also abundant in gardens

The 2-lobed fruits readily distinguish this species. The genus name comes from the Greek words meaning *crown foot*, referring to the deeply cleft leaves. *Didymus* refers to the globular fruits and means *twins* or *in pairs.* This species has the feathery foliage of good treading weeds; the finely dissected leaves resist damage when trampled. In untrampled areas, plants grow more upright and are larger. The resemblance of the fruits to warts gives this plant the common name of *Wart Cress.*

Cruciferae (Mustard family)

Euphorbia peplus

Petty Spurge

Annual, introduced from Europe

Habit: erect, slender, yellow-green stems, 1-2 dm. tall

Flowers: inconspicuous, greenish, surrounded by crescent-shaped glands (Jan.-Aug.)

Leaves: upper leaves oval, lower leaves broadly ovate to round, with short petioles

Fruits: globular, lobed capsules with 2 crests on the back of each lobe

Habitat: moist shady places in garden beds, yards

Spurge comes from the Old French word meaning *the purging herb*, the seeds having been used as a purgative. The milky white sap is characteristic of all members of this family. The seeds of *Euphorbia* species (although not usually this one) often have an attached fat body that is attractive to ants, and therefore aids in seed dispersal. This species has separate male and female flowers on the same plant (monoecious).

Euphorbiaceae (Spurge family)

Glossary

achene-a dry one-seeded fruit not splitting at maturity

auricle-an ear-shaped lobe or appendage

axil-angle between leaf and stem (see p. *ix*)

bract-a rudimentary leaf near a flower or inflorescence

calyx-all of the sepals together (see p. *ix*)

capsule-a dry fruit with few to many seeds, splitting at maturity

corolla-all of the petals together (see p. *ix*)

cyme-a branched inflorescence in which the central flower opens first

cymose-cyme-like

dehiscent-splitting open at maturity

disk flower-tubular flower in a Sunflower family infloresence (see p. *x*)

drupe-a fleshy, one-seeded fruit not splitting at maturity

drupelet-a small drupe, such as the small parts of a blackberry fruit

follicle-a dry fruit that splits down one side at maturity

head-a dense cluster of sessile flowers or fruits (see p. *x*)

indehiscent-not splitting apart at maturity

inflorescence-flower cluster

irregular-having differently shaped petals or sepals

lanceolate-lanceshaped, several times longer than wide

mericarp-a portion of a fruit that splits into parts

merous-parted, having sections

node-the joint on the stem where leaves, branches, or flowers arise (see p. *ix*)

nutlet-a small, one-seeded fruit with a hard wall

oblanceolate-inversely lanceolate

obovate-inversely ovate

ovary-seed-containing part of the flower (see p. *ix*)

ovate-egg-shaped in outline

palmate-having lobes or divisions radiating from a central point

panicle-a branched inflorescence

pappus-scales, or bristles attached to an achene (see p. *x*)

pedicel-the stalk to a single flower (see p. *ix*)

petal-a usually brightly colored flower part between the sepals and stamens (see p. *ix*)

petiole-the stalk to a leaf (see p. *ix*)

pinnate-opposite leaflets or lobes on a central axis

raceme-an unbranched inflorescence with stalked flowers

ray flower-the strap-shaped flower in a Sunflower family inflorescence (see p. *x*)

receptacle-the expanded portion of the flower stalk that bears the flower(s) (see p. *x*)

rosette-a dense basal cluster of leaves

schizocarp-a dry fruit that splits into 2 or more parts at maturity *sepal*-a usually green part of the outer whorl of flower parts (see p. *ix*)

sessile-without a stalk, borne directly on stem (see p. *ix*)

silicle-a dry dehiscent fruit short and squat in shape (Mustard family)

silique-a dry dehiscent long and narrow in shape (Mustard family)

simple-not divided into separate segments; may be lobed

spatulate-broad and rounded at tip and tapering towards the base; spatula-shaped

spike-an inflorescence made up of sessile flowers on a long unbranched stem

stamen-the pollen-producing part of the flower (see p. *ix*)

stigma-the part of the flower receptive to pollen (see p. *ix*)

stipule-a bract-like appendage at the base of the petiole or leaf (see p. *ix*)

style-stalk-like structure connecting ovary and stigma (see p. *ix*)

umbel-a flat-topped inflorescence with all pedicels attached at the same point

Index

Additional References

Balls, Edward K. 1962. *Early uses of California plants.* Berkeley: University of California Press.

Clarke, Charlotte B. 1977. *Edible and useful plants of California.* Berkeley: University of California Press.

Harrington, H.D. and L.W. Durrell. 1957. *How to identify plants.* Chicago: Swallow Press.

McHoul, Lilian and Celia Elke. 1979. *Wild flowers of Marin: a layman's handbook.* Fairfax: Tamal Land Press.

Munz, Philip A. and David D. Keck. 1968. *A California flora and supplement.* Berkeley: University of California Press.

Niehaus, Theodore F. and Charles L. Ripper. 1976. *A field guide to Pacific States wildflowers.* Boston: Houghton Mifflin Co.

Ornduff, Robert. 1974. *An introduction to California plant life.* Berkeley: University of California Press.

Sharsmith, Helen K. 1965. *Spring wildflowers of the San Francisco Bay Region.* Berkeley: University of California Press.

Additional copies of this book may be obtained from your local bookseller, or by ordering direct, adding 10% postage, plus applicable sales tax.

Appletree Press
2644 Dwight Way #6
Berkeley, California 94704